Carolina Rosai Mendes
Guilherme Dilarri
Renato Nallin Montagnolli

Degradation and Biosorption of Textile Dye in a Yeast Bioreactor

Carolina Rosai Mendes
Guilherme Dilarri
Renato Nallin Montagnolli

Degradation and Biosorption of Textile Dye in a Yeast Bioreactor

A study with Saccharomyces cerevisiae

ScienciaScripts

Imprint
Any brand names and product names mentioned in this book are subject to trademark, brand or patent protection and are trademarks or registered trademarks of their respective holders. The use of brand names, product names, common names, trade names, product descriptions etc. even without a particular marking in this work is in no way to be construed to mean that such names may be regarded as unrestricted in respect of trademark and brand protection legislation and could thus be used by anyone.

Cover image: www.ingimage.com

This book is a translation from the original published under ISBN 978-613-9-62105-7.

Publisher:
Sciencia Scripts
is a trademark of
Dodo Books Indian Ocean Ltd. and OmniScriptum S.R.L publishing group

120 High Road, East Finchley, London, N2 9ED, United Kingdom
Str. Armeneasca 28/1, office 1, Chisinau MD-2012, Republic of Moldova, Europe
Printed at: see last page
ISBN: 978-620-7-71326-4

CONTENTS

SUMMARY

Textile effluents release large quantities of dye residues into the environment, polluting and altering the entire aquatic ecosystem. In addition to visual pollution, the compounds are highly toxic and are considered carcinogenic, mutagenic and bioaccumulative in biota.

A number of methods are used to treat these textile waste compounds. However, some methods have economic disadvantages and others are unfeasible for the textile industry due to their time-consuming nature.

The method using microorganisms has shown promise with the use of some fungi from the biodegradation and biosorption of waste.

Biodegradation allows the toxic compounds present in the effluent to be broken down into organic compounds that are less aggressive to the environment.

Biosorption removes these toxic compounds by binding the waste molecule to the cell wall of the microorganism through chemical bonds, when the pH values are adjusted for the best adsorbent/adsorbate affinity.

A study was therefore carried out on the removal of the reactive dye Solar Orange 2GL P250, using the commonly used yeast *Saccharomyces cerevisiae* for biosorption and biodegradation analyses in a pilot-scale bioreactor system, in order to minimize environmental water impacts.

The biodegradation studies made it impossible to quantify the decolorization of the dye, due to the alteration of the initial molecule, which generated a new color and not a decrease in color. Furthermore, the 72-hour treatment time is not feasible for the textile industry.

The studies showed significant results, even at low biomass concentrations, in biosorption removal in a short treatment period, which makes it viable for the textile industry. Thus, the use of *S. cerevisiae* biomass is a potential microorganism for the biosorption treatment of textile effluents containing surplus reactive dyes.

INTRODUCTION

Environmental problems are increasingly evident in contemporary society and are observed through the various physical, chemical and biological changes in the quality of soil, air and water.

One of the great problems of the century is the quality of the water body, which in many areas exceeds the standards established as tolerable.

Textile industrial activities consume a lot of water in their processes, generating a high volume of effluents and consequently contributing to increased levels of contaminants in natural waters (SALLES *et al.*, 2006). This fact, coupled with the low use of inputs, especially dyes, means that the textile industry is responsible for generating waste with a high organic load and strong coloration (KUNZ *et al.,* 2002).

It is estimated that 10-15% of the total load of dyes enters the environment through industrial effluents (REVANKAR *et al.*, 2007).

The organic load released alters the ecosystem due to increased turbidity in the water, hinders the penetration of solar radiation, which generates changes in photosynthetic activity and the solubility regime of gases (SOUZA *et al.*, 2005). Furthermore, the disposal of textile wastewater poses a danger of bioaccumulation and eco-toxicity risks to the environment (CONTRERAS *et al.*, 2012; HASSAN et *al.,* 2014; ZHU et *al.,* 2013), due to dyes and toxic compounds such as benzene, toluene and ethyl-benzene, naphthalene, anthracene and xylene (GONG *et al.,* 2007).

These compounds can remain in terrestrial and aquatic biota for around 50 years, jeopardizing the stability of these ecosystems and the life around them (GUARATINI *et al.*, 2000).

The total production of dyes in the world is estimated at 800,000 tons per year (GANODERMAIERI *et al.*, 2005). In Brazil alone, it is estimated that more than 20,000 tons of these products are consumed annually by textile industries (DALLAGO *et al.*, 2005).

Among the classes of textile dyes currently available are acid, basic, disperse, azo, diazo, metallized and reactive dyes, which are widely used in industries and already account for 60% of world consumption (CASTANHO *et al.*, 2006).

In general, the color of a dye is the result of the interaction of chromophore functional groups between azo functions (-N=N-) and aromatic species (GALINDO *et al.*, 1999).

The functional groups, called auxochromes, in reactive dyes can be chlorine, hydroxyl, bromine, nitro, sulfonic, amino, methoxy, ethoxy and ethyl (KIMURA *et al.*, 2000), which allow them to reach different wave frequencies in the visible spectrum, and therefore different shades of color through hypsochromic and batochromic effects. In addition, the auxochrome groups determine the anionic character of the dye.

Reactive dyes contain an electrophilic (reactive) group capable of forming covalent bonds with anionic and cationic interactions with the hydroxyl groups of cellulose fibers, with the amino, hydroxyl and thiol groups of protein fibers and also with the amino groups of polyamides (GUARATINI *et al.*, 2000).

These bonds ensure greater stability between the dye molecules and the fabric fibers after the dyeing processes and the surplus goes into the form of effluent from the industry.

CONAMAM Resolution N°357/2005 allows residual disposal in accordance with regulations and resolutions issued by the Ministry of the Environment, so that the effluent does not influence the color of the water body when incorporated into it.

However, effluents do not always meet these standards, due to the treatment method used by the industry and the degree of discoloration of the dye molecule.

Due to the resulting environmental problems, many studies associated with new technologies have been proposed for the treatment of textile effluents, using physical-chemical and/or microbiological processes, such as ozonation (SANTOS *et al.*, 2011), photo-electrochemical degradation (BRUNELLI *et al.*, 2009), subcritical flow (HOSSEINI *et al.*, 2010), biological degradation (KAMIDA et al., 2005), electrodialysis (SALEHIZADEH et *al.*, 2003), adsorption (EL-KHAIARY *et al.*, 2009) and biosorption (MÓDENES, *et al.*, 2011).

Recently, many studies have shown that the use of microorganisms can be considered a valid alternative to physico-chemical methods and/or enzymatic or microbiological degradation (PRIGIONE *et al.*, 2008).

Textile dye decolorization processes using biological methods occur in two ways: by biosorption or by biodegradation (OGUGBUE et al., 2011).

Biodegradation can be expressed as the way in which a microorganism, through oxidation, reduction, hydroxylation, dehalogenation, hydrolysis and demethylation reactions, begins to break down the molecule, making it susceptible to the action of other organisms (RAJAGURU et al., 2000).

Thus, the action of these enzymes results in the decomposition of compounds into commonly occurring organic and inorganic products such as water and carbon dioxide, and toxic metabolites may be formed when degradation is incomplete (DELLAMATRICE *et al.*, 2005).

Although there is a huge range of enzymes, only the oxy-reductive ones are capable of completely degrading dyes (CHACKO *et al.*, 2011).

When degradation is incomplete, dyes can undergo transformations in their molecule that lead to the formation of carcinogenic substances such as amines and free metals (MITTER *et al.*, 2012).

A certain group of microorganisms may be able to degrade one product, but not another that is structurally different, and there is specificity between the microorganism and the compound (GOSZCZYNSKI *et al.*, 1994).

However, the treatment of textile dyes by complete degradation guarantees the transformation of recalcitrant toxic compounds into compounds that are not toxic to the environment.

The biosorption method can be defined as a binding of the solute to microbial biomass such as bacteria, fungi and algae, through a process that does not involve metabolic energy transport, although each process can occur simultaneously when living biomass is used (KAUSHIK *et al.*, 2009).

In addition to providing recognized advantages, such as biomass can be reused; compounds can be removed from the solution regardless of the degree of toxicity; operating times are short when equilibrium is reached; it does not produce secondary compounds with high toxicity and can be highly selective (KAC *et al.*, 2002; JIANLONG *et al.*, 2001); the adsorbed material can be reused; in addition to being economically viable (FAVERE *et al.*, 2010).

However, the biosorption capacity depends on the type of biomass, the type of adsorbate, the presence of competing ions in the medium and pH (MITTER *et al.*, 2012). Furthermore, the residual dye is only transferred by mechanical removal from a liquid to a solid state.

Among the various types of biomass researched, fungi have been shown to be particularly suitable for the process of textile effluent treatment by dye biosorption (PRIGIONE *et al.*, 2008; AKSU et *al.*, 2000; YESILADA *et al..*, 1995), due to their selective retention capacity by ion exchange (charged sites on the adsorbent) (VOLESKY *et al.*, 1999) which are associated with the cell wall of the microorganism via adsorption without degradation (SUMATHI *et al.*, 2000).

Fungal biomass is also widely used in degradation treatment, playing an important role in the production of enzymes such as peroxidase, phenoloxidase and lacases (VITOR *et al.*, 2008), which promote the complete degradation of the coloring compound.

The yeast *Saccharomyces cerevisiae*, also known as baker's yeast, is a facultative aerobic microorganism, i.e. it has the ability to adjust metabolically in both aerobic and anaerobic conditions (DOS SANTOS *et al.*, 2010).

From an industrial point of view, *S. cerevisiae* is one of the most common biocatalysts, mainly due to its easy availability, low cost, not requiring special instruments, easy handling, non-pathogenicity, efficiency compared to conventional catalysts (BARALDI *et al.*, 2004).

This study aims to evaluate the biosorption and biodegradation potential of the Solar Orange 2GL P250 commercial dye of the reactive group, widely used in the textile sector, containing sulfonic groups by *S. cerevisiae* biomass in a pilot-scale bioreactor, in order to minimize environmental water impacts.

OBJECTIVE

The study aims to evaluate the biosorption and biodegradation potential of the Solar Orange 2GL P250 dye using *S. cerevisiae* biomass in a pilot-scale bioreactor using UV-Vis spectrometry.

MATERIALS AND METHODS

Dye.

The Solar Orange 2GL P250 dye (Figure 1), which belongs to the group of sulphonating reagents, has a Chemical Abstract Service (CAS) database registration number of 1325-54-8, a molecular weight of 299 g.mol^{-1} and an empirical formula of $C_{12}H_{10}N_3NaO_3S$, was studied without prior purification.

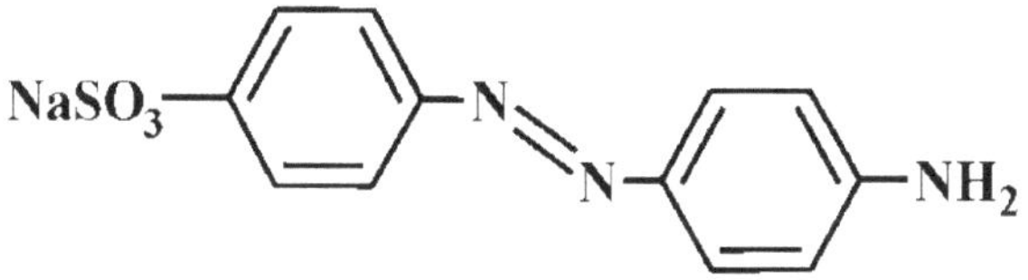

Figure 1. Structural formula of the reactive dye Solar Orange 2GL P250.

Saccharomyces cerevisiae.

The yeast _S. cerevisiae_ was used in this study from biological yeast marketed in freeze-dried form by Pak Gida Uretim ve Pazarlama A. S. Cd. No: 5/6 34349, Lot: 369366 14. Weighings of 500; 800; 1000 mg were made for the biosorption tests in a bench study.

Study of pH value variation in relation to the Batochromic and Hypsochromic effect of the dye.

In the pH variation study, the Solar Orange 2GL P250 dye solutions were evaluated starting at a concentration of 100 mg/L in order to measure the variations in the dye's wavelength.

The pH values were changed by adding solutions of 0.1 mol/L hydrochloric acid (HCl) and 0.1 mol/L sodium hydroxide (NaOH).

The pH values were only possible with acidic and basic solutions due to the presence of the dye, which worked in this study as a buffering agent for the solution, as verified experimentally.

A bench pH meter (Tecnopon, model MPA-210) was used to monitor the pHs of the dye solutions, set at pH = 2.8; pH = 5.2; pH = 6.6 and pH = 9.3.

In the study, the effect of pH values on color changes was verified by the effects of bathochromic displacement (displacement of the absorption band at the longer wavelength within the visible spectrum) or hypsochromic displacement (displacement of the adsorption band at the shorter wavelength within the visible spectrum).

Analyses at the different pH values were carried out using quartz cuvettes and an Ultraviolet-Visible spectrophotometer (Thermo Scientific, model BioMate 3S), scanning between 300 and 800 nm.

Calibration curve for the Solar Orange 2GL P250 dye.

In order to quantify the concentration of the dye remaining in solution in the biosorption studies, an initial standard solution of 100 mg/L was prepared, from which secondary dilutions were made, with final concentrations of 0.5 mg/L, 1 mg/L, 2 mg/L, 5 mg/L, 10 mg/L, 20 mg/L, 30 mg/L and 40 mg/L. Scans were carried out at wavelengths of 300 to 800 nm in a UV-Vis spectrophotometer for each concentration of the dye, using a quartz cuvette with an optical path of 5 mm.

The calibration curve was obtained using different concentrations of the dye as a function of absorbance at maximum wavelength in UV-Vis spectrophotometer analyses.

From the absorbance values at the maximum wave peak, it was possible to obtain a graph of the standard line in the Origin 6.0 program for the values of A, B and R in equation 1.

$$Y = A + B.X$$

Equation 1. equation of the line

Bench study of the Solar Orange 2GL P250 dye.

The biosorption tests of the Solar Orange 2GL P250 dye solution were carried out using different concentrations of freeze-dried _S. cerevisiae_ biomass.

The analyses were carried out according to the contact time of 120 minutes, for the equilibrium of the system in a bench study at room temperature of $25^{+/-} 2°C$.

The study was carried out using the mother solution of the dye at a molar concentration of 16.72×10^{-2} mmol/L (50 mg/L) at different hydrogen potentials, determined after studying the variation in pH values, with the possible values of pH = 4; pH = 5; pH = 6.

Variations of biomass were simultaneously added as an adsorbent agent at 500 mg; 800 mg; 1000 mg in Beakers containing 30 mL of the dye solution and null as a control.

Aliquots were taken from each sample after the equilibration period and then centrifuged (CentriBio centrifuge) at a speed of 4,000 rpm for 15 minutes.

Thus, the supernatant tends to separate by gravity from the liquid, so as not to generate interference (noise) in the UV-Vis spectrophotometer analysis, adjusted to a wave frequency for scanning between 300 and 800 nm.

The experiment was analyzed from the isotherm graph, using the models of the Langmuir equation (Equation 2) and its linearization (Equation 3), Freundlich (Equation 4) and its linearization (Equation 5), in order to calculate the maximum capacity (q_m) of biosorption of the cell wall of *S. cerevisiae* at pH = 4, pH = 5 and pH = 6 .

$$q = \frac{Kqm.\,Ce}{1 + KCe}$$

Equation 2.

$$\frac{Ce}{q} = \frac{Ce}{qm} + \frac{1}{Kqm}$$

Equation 3.

$$n.\,Log\,Ce = Log\left(\frac{x}{m}\right) - Log\,K$$

Equation 4.

$$Log\left(\frac{x}{m}\right) = Log\,K + n.\,Log\,Ce$$

Equation 5.

For Langmuir (q) is the concentration of the dye; (qm) is the maximum biosorption capacity; (K) is the equilibrium constant related to the free energy of biosorption which corresponds to the affinity between the surface of the yeast cell wall and the solute and (Ce) is the equilibrium concentration.

For Freundlich (n) is the slope, (Log K) is the intercept in (Y) and (Ce) is the equilibrium concentration.

Biosorption kinetics of the dye Solar Orange 2GL P250.

To do this, an initial solution of 50 mg/L of the dye was prepared and distributed in three Beakers with equal volumes of 30 mL.

In each, 500 mg, 800 mg and 1000 mg of freeze-dried *S. cerevisiae* biomass were added at pH 6 (characteristic of the dye in solution).

Determinations of the dye in the solution were carried out at sequential intervals of 15 minutes over 105 minutes.

The analyses were evaluated by UV-Vis spectrophotometry with absorbance values adjusted according to the calibration curve by quantifying the dye in mg/L.

Study in a pilot-scale bioreactor.

In order to evaluate the treatment by biosorption and biodegradation of textile effluent residues, a pilot-scale vertical bioreactor made of tempered glass was used.

The drum has a volume of 2000 mL, oxygenation by an integrated aeration pump at 1 vvm which provides the aerobic organism with oxygen (the final electron acceptor) and a controlled temperature of 30°C, which is conducive to the growth of the *S. cerevisiae* yeast.

The study made it possible to verify the decolorization potential of the reactive contaminant Solar Orange 2GL P250 on a pilot scale by *S. cerevisiae.*

Therefore, the dye solution at a concentration of 50 mg/L for a volume of 2000 mL did not change in pH value, remaining at pH = 6, characteristic of the dye in solution. After thermal equilibrium in the bioreactor (30°C), 3.3×10^4 mg of freeze-dried *S. cerevisiae* biomass was added in proportion to the 500 mg bench study. After 90 minutes, aliquots were taken and analyzed in a UV-Vis spectrophotometer set to scan between 300 and 800 nm.

For biodegradation analysis, aliquots were taken after 24, 48 and 72 hours.

FT-IR (Fourier Transform Infrared) Spectrometer Analysis

Analyses of the dye and the adsorbent material before and after adsorption at pH 2.5 and 8.5 were carried out using Fourier transform infrared spectra, set to scan between 400 and 4000 cm^{-1} on a Shimadzu model 8300 spectrophotometer. The readings were taken by making a pellet containing 149 mg of KBr and 1 mg of sample of each crushed spherical material.

The tablet material was subjected to a compression of 40 KN for approximately 5 minutes. The tablets were then placed in the FT-IR for absorbance readings with a resolution of 4 cm^{-1} and 32 scans.

Acute toxicity test.

For the acute toxicity test, a species of freshwater microcrustacean *Daphnia similis* was used in its young phase (6 to 24 hours old).

The organisms in the growth phase remained in a solution at pH 6.50 in a temperature-controlled incubator at 20° C, with 1000 Lux lighting and a photoperiod of 16 hours of light and 8 hours of darkness.

The feed during this period was a 2 mL suspension of the microalga *Pseudokirchneriella subcapitata* at a concentration of 3.00 ± 3. 50×107 mL

(determined by Petroff-Hausser microscopic counting) and 1 mg of fish feed.

D. similis was exposed to different concentrations of the dye Solar Orange 2GL P250 for a period of 48 hours to determine the average lethal concentration (CL_{50}) and the maximum lethal concentration (LC_{100}) determined by the Trimmed Spearman-Karber method.

A sample of known concentration of sodium dodecyl sulfate ($NaC_{12}H_{25}SO_4$) was used as a control.

The toxicity test was also carried out after adsorbent treatment with QC beads. Thus, 10, 20 and 30 mg of the adsorbent material were added to 100 mL of dye at the same LC50 concentration at pH 6.50 at the equilibrium time as determined in the kinetic study.

The adsorbent material was separated from the solution and *D. similis* remained in contact with the solutions for 48 h at 20°C. The number of living organisms was then counted.

The tests were carried out with three repetitions and submitted to non-parametric statistical data using the Kruskal-Wallis method in the BioEstat 4 program.

Results and discussion

In the pH value variation study, the Solar Orange 2GL P250 reactive dye solution showed no absorbance variations and/or wavelength shifts characteristic of bathochromic or hypsochromic effects in UV-Vis spectrophotometry analyses as a function of pH value.

In this way, it was possible to quantify the dye for the calibration curve without using a buffer solution to adjust the pH value.

The maximum wavelength of the dye solution was set at 413 λ (Figure 2).

In addition, the dye showed intrinsic buffering characteristics in solution, due to the amino (NH_2) and azo (N=N) functional groups present in the dye molecule, which gives it a pH of 6.2 in solution.

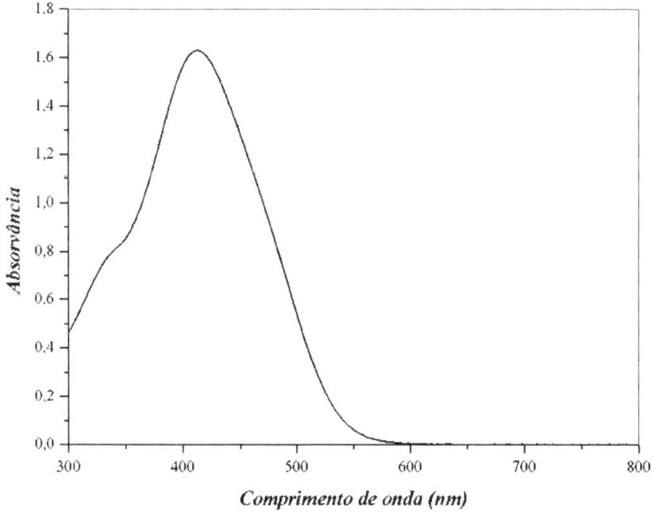

Figure 2 - UV-Vis spectrum of the characteristic wavelength of the Solar Orange 2GL P250 dye in

aqueous medium at pH = 6.

The concentrations of dye remaining in the solutions were calculated using the calibration curve from 0 to 40 mg/L, shown in figure 3.

From the curve, parameters such as linearization, angular coefficient and linear coefficient were evaluated according to the straight line equation using Origin 6.0 software.

The equation of the line $Y = A + B.X$, where the linear coefficient $A = 0.0038$; angular coefficient $B = 0.03361$; Y will be the absorbance readings for the isotherm and kinetics studies; and for X the concentration values.

With regard to the linear correlation coefficient, the value was R= 0.99998, which indicates an excellent linear fit in relation to the points on the calibration curve.

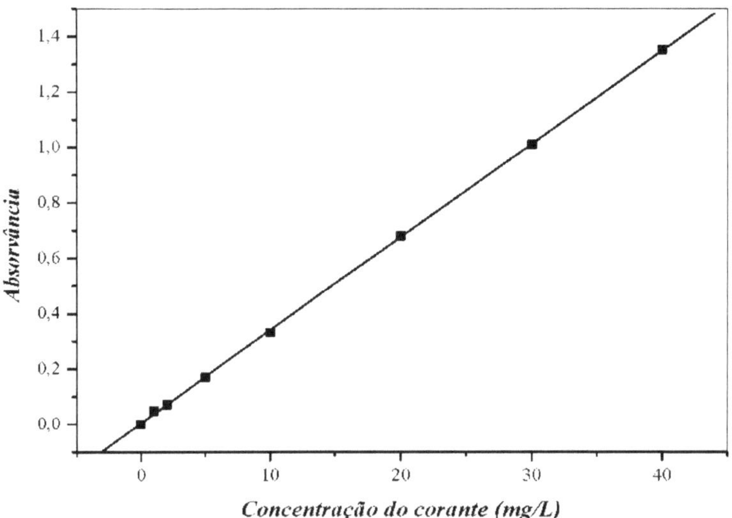

Figure 3: Calibration curve for the Solar Orange 2GL P250 dye

Figure 4 shows the solutions used to make the calibration curve for the Solar Orange 2GL P250 reactive dye.

Figure 4.   Preparation solutions for the calibration curve of the Solar Orange 2GL P250 dye.

The bench studies showed isotherms for the biosorption of the dye with more efficient results at pH = 4 than at pH = 5 and pH = 6 (Figure 5).

This is due to the characteristics of the *S. cerevisiae* yeast cell wall and the dye in relation to its functional groups, as described by Mitter et al. (2012).

At pH = 4, the functional groups of the cell wall have more active sites available ($-NH_3^+$), due to the degree of acidification, than at pH = 5 and pH = 6. Thus, when comparing pH = 4, the best performance was obtained; however, pH = 5 and pH = 6 also obtained dye removal results.

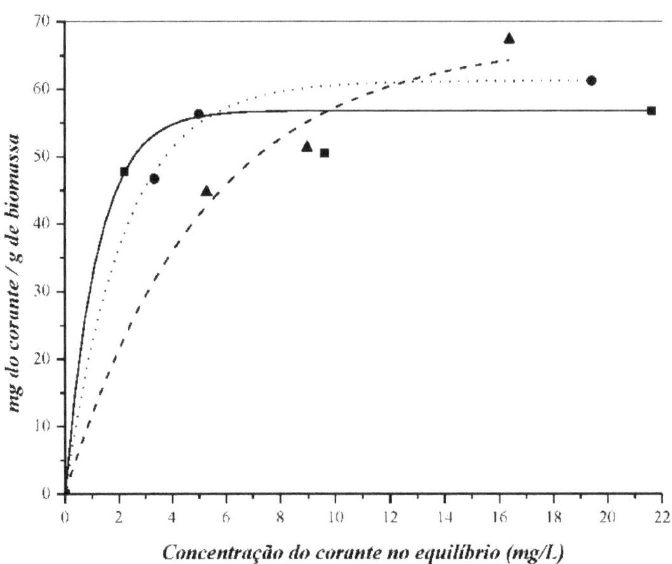

Figure 5 Isotherms of the dye Solar Orange 2GL P250 at different values of pH = 4 (■), pH = 5 (·

), pH = 6 (A), during the bench study.

Figure 6 shows the discoloration of the solution after contact with the adsorbent and varying the mass of yeast compared to the initial control.

The discoloration of the solution is proportional to the increase in the mass of the adsorbent.

It is possible to state that the total removal of the dye molecule in solution can be easily achieved just by changing the concentration of the yeast mass and the pH value.

22

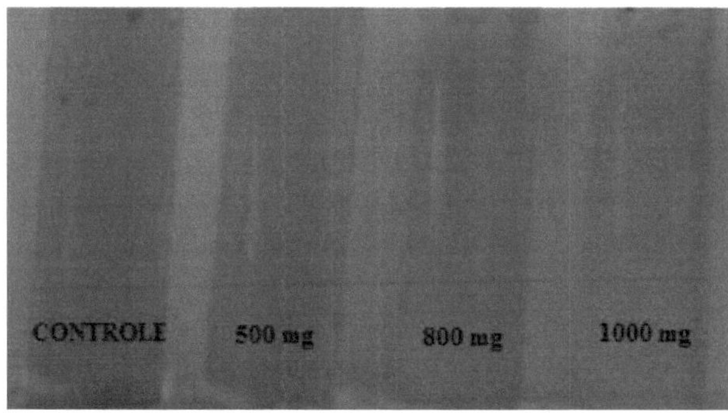

Figure 6.   Discoloration by adsorption process varying the mass of adsorbent.

The linearization of the Langmuir and Freundlich isotherms made it possible to determine the maximum capacity (qm) of the biosorption studies with variations in pH values.

It also makes it possible to determine the effect of the adsorbate/adsorbent interaction, analyzing adsorption through monolayers or multilayers.

Each pH value presented a pattern of relative correlation coefficients (R) for each isotherm, as shown in Table 1.

Table 1. Relative correlation coefficient (R) for isotherm analysis.

	Freundlich	*Langmuir*
pH		
4.0	R = 0,95206	R = 0,99793
5.0	R = 0,83100	R = 0,99976
6.0	R = 0,98045	R = 0,97073

Studies on the composition of the *S. cerevisiae* cell wall have revealed the presence of glucan (48 - 60%), which is a polymer of glucose units with $\beta(1$-$3)$ and $\beta(1$-$6)$

bonds, mannoproteins (20 - 23%), chitin (0.6 - 2.7%), which is composed of β(1-4) N-acetylglucosamine and a small portion of lipids (FLEET *et al.*, 1985).

The distribution of these components is organized into two main layers, the outer one composed of mannoproteins and the inner one of glucan and chitin, in a structure interconnected by covalent bonds (HA *et al.*, 2002).

In general, the cell wall of *S. cerevisiae* yeast is considered to be the most prevalent site for binding dyes, such as the chemical groups acetamido, amido, fosfato, amino, amina, sulfidrila, carboxila and hidroxila (VOLESKY *et al.*, 1995).

The Solar Orange 2GL P250 dye has the intrinsic characteristics of reactive sulphonic groups, which makes it soluble in aqueous media.

In solution, the sodium ion (Na^+) dissociates and leads to the appearance of the sulfonic group ($-SO_3^-$), which is capable of forming ionic bonds, such as with amino groups.

In this way, possible mechanisms of interaction between the dye and the chitin found in the cell wall of the yeast *S. cerevisiae are* argued, respecting the Langmuir and Freundlich isotherms.

As shown in the isotherms at pH = 4 and pH = 5, the results demonstrate monolayer biosorption, respecting the Langmuir isotherm equation.

In this way, in acidified media, it is proposed that the yeast cell wall increases the amount of active sites available for binding the dye, due to the protonated amino groups ($-NH_3^+$) which promote the emergence of strong interactions with the basic site of the sulphonic groups ($-SO_3$) of the dye (Figure 7).

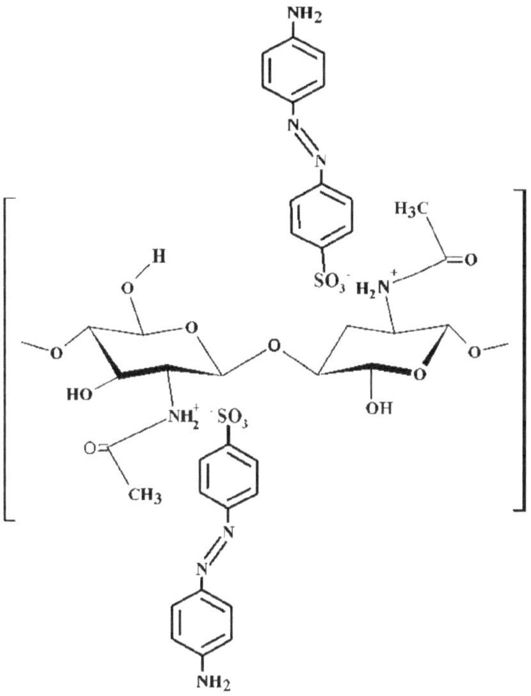

Figura 7.   Possible hydrogen bond between the amino group of chitin present in the cell wall of *S. cerevisiae* and the sulfonic group of the dye molecule.

Another proposed mechanism also respects the Langmuir isotherm equation, being monolayer and can occur through hydrogen bonds between the sulfonic groups ($-SO_3^-$) and the hydroxy groups ($-OH_2^+$) present in the chitin of the yeast cell wall (Figure 8).

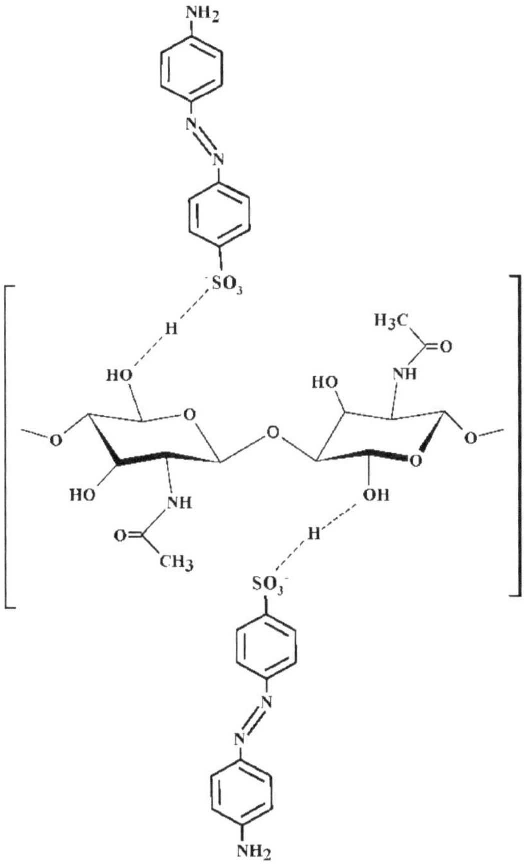

Figura 8.   Possible hydrogen bond between the hydroxy group of chitin present in the cell wall of *S. cerevisiae* and the sulfonic group of the dye molecule

At pH = 6, the amino groups ($-NH_2^+$) of chitin are less protonated and the dye can bind to the yeast cell wall through hydrogen bridges.

Therefore, biosorption respects the Freundlich isotherm equation, being a multi-layered biosorption of interactions gathered around the cell wall of the microorganism.

This may be due to the hydrogen bonds between the dye and the chitin, as well as the interactions of the dye molecules with other dye molecules, through the hydrogen bonding of the azo group (N=N) with the amino group (-NH2) shown in Figure 9.

This type of bond is weaker than the covalent bonds that occur at pH = 4 and pH = 5.

Thus, the biosorption of the dye by the yeast at pH = 6 is lower than at pH = 4 and pH = 5. However, the acidic pH value is more aggressive when disposed of in the environment than pH values close to neutral.

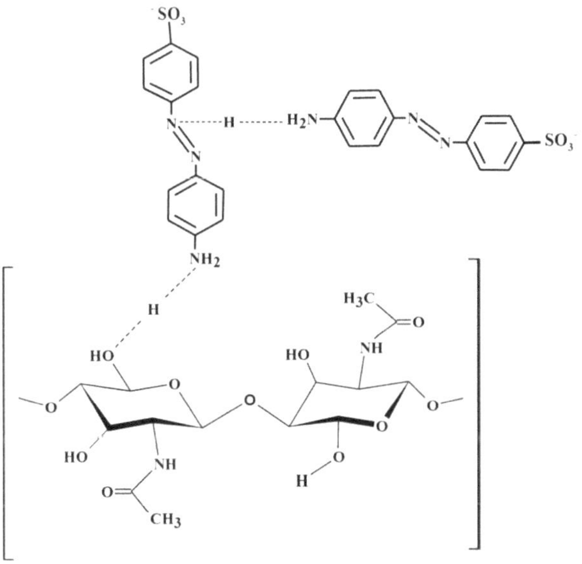

Figure 9. Possible ionic bond between the azo group (N=N) in the dye molecule and the sulfonic group (so₃⁻) in the dye molecule.

In the kinetic study, the minimum time needed for the biomass/dye system to reach equilibrium conditions was observed.

Figure 10 shows the kinetic evolution of dye removal as a function of time. For 1000 mg and 800 mg of *S. cerevisiae* biomass, the system reaches equilibrium after 60 minutes; for 500 mg of biomass, the system reaches equilibrium after 90 minutes.

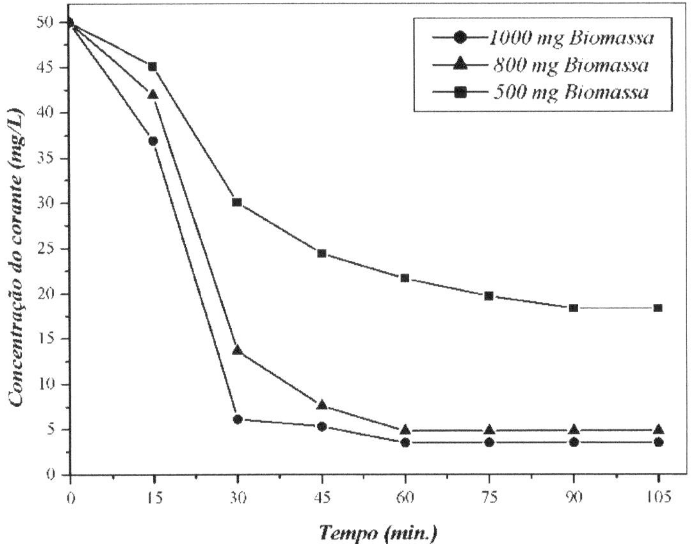

Figura 10.   Biosorption kinetics of the reactive dye Solar Orange 2GL P250 at pH = 6.

The decrease in absorbance values at the maximum wave peak during the decolorization of the Solar Orange 2GL P250 dye occurred through biosorption and/or biodegradation.

When biosorption occurs, the absorbance values at the maximum peak decrease without changes in length or wave shifts.

Figure 11 shows decolorization spectra by biosorption in a pilot bioreactor system, by reducing the concentration of the dye in solution at pH = 6 (pH characteristic of the dye in solution).

28

When adsorbed, the dye molecules, previously dispersed, are immobilized in the cell wall, which leads to their gradual mechanical removal from the solution until equilibrium is reached (90 minutes, as described in the kinetic study).

Thus, the spectrum shows a decrease in absorbance with no change in the characteristic wavelength at 413 λ.

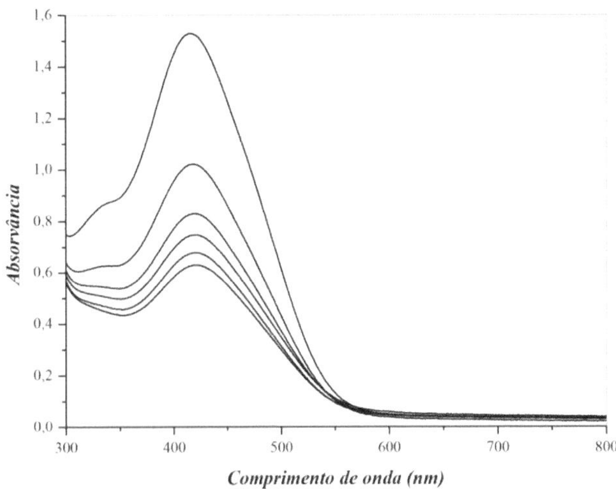

Figura 11. UV-Vis spectra of the decolorization by biosorption of the Solar Orange 2GL P250 dye by *S. cerevisiae* in a pilot bioreactor system.

When biodegradation takes place, the maximum wave peaks decrease in different proportions, forming new peaks at different wavelengths in the UV-Vis spectrum.

This is due to the resulting absorbance of new metabolites or fragments of the dye molecule after degradation (DANESHVAR *et al.,* 2007). Figure 12 shows the spectra of the biodegradation of the dye at 24, 48 and 72 hour intervals, through the formation of new peaks in the wavelength in solution at pH = 6.

In addition, changes in the plasmon bands only occurred after 24 hours in the pilot bioreactor; prior to this period, there was only a decrease in the absorbance values due to the occurrence of biosorption. In general, degradation makes it impossible to quantify the remaining dye in solution compared to the control, since the molecule is no longer the same with changes in spectral structure and changes in chromophore groups due to the change in color.

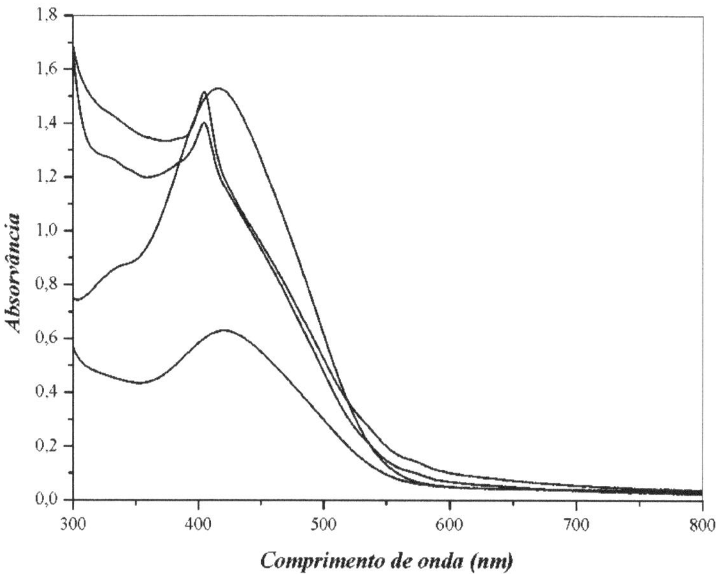

Figura 12. UV-Vis spectra of the decolorization of Solar Orange 2GL P250 dye by *S. cerevisiae* biodegradation in a bioreactor system.

With regard to biodegradation, a possible mechanism for breaking the primary bond was proposed, by breaking the azo group (N=N) of the Solar Orange 2GL P250 dye molecule (Figure 13).

This is possible due to the intrinsic characteristics of the π-bonded group, which makes it easier to break because it is more unstable due to the planar region than in other regions of the molecule.

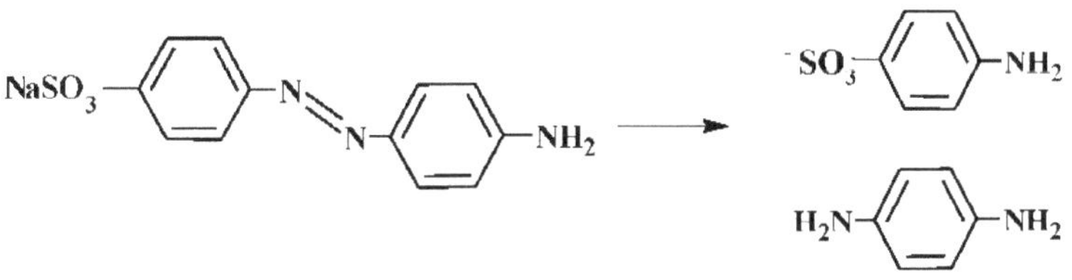

Figura 13. Possible molecules formed from the Solar Orange 2GL dye in biodegradation processes by *S. cerevisiae*.

The effectiveness of color removal can be assessed by a spectrophotometrically permissible standard defined in the literature (COOPER, 1993).

In general, it is possible to assess the degree of contamination through the absorbance of the residue in solution, with an absorbance of up to 0.06 being tolerable.

The study in a pilot bioreactor (Figure 14) showed, based on UV-Vis spectrophotometer analysis, a 70% reduction in coloration with an absorbance of 0.42, using a pH value of 6 in solution, with an increase of 3.3×10^4 mg of yeast biomass (this weight refers to 500 mg of biomass) in a maximum period of 90 minutes for biosorption (Figure 15).

However, the use of biomass is directly proportional to the reduction in staining, thus guaranteeing better results when compared to the permitted absorbance value.

Figura 14. Pilot-scale bioreactor for the biodegradation of Solar Orange 2GL P250 by *S. cerevisiae*

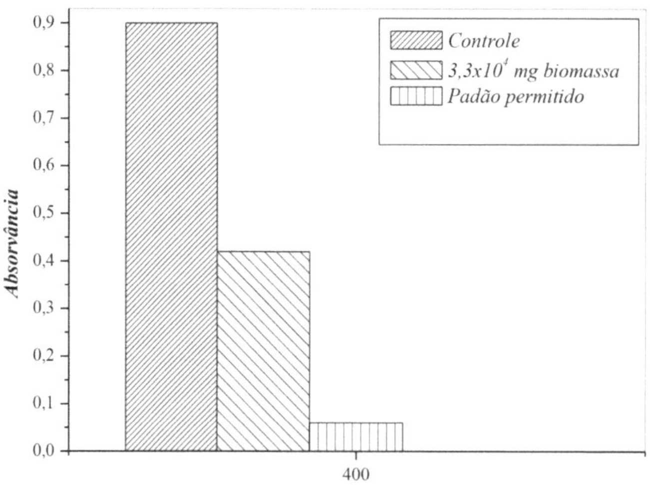

 Graph of the biosorption data of the dye Solar Orange 2GL P250 by *S. cerevisiae* compared to the standard allowed by CONAMA.

In the toxicity studies, the microcrustacean *D. similis* showed considerable sensitivity to the dye Solar Orange 2GL P250, with LC_{100} values of 1100 µg mL^{-1} and CL_{50} values of 200 µg mL^{-1} .

The results showed that yeast biomass reduced the toxicity of the dye in the solution (Table 2).

Despite the lower mortality of microcrustaceans after removing the dye, this reduction was not statistically significant.

Using 500 mg of these adsorbent materials, it was possible to reduce microcrustacean mortality.

This demonstrates the efficiency of this material in removing the toxicity of the dye solution after adsorption.

One of the probable reasons for the lack of results regarding the reduction in *D. similis* mortality using *S. cerevisiae* biomass was the pH value and temperature used in this test, as they were not ideal in relation to the adsorption efficiency shown in the isotherm studies.

Table 2. Toxicity data with *D. similis*.

Yeast	Initial concentration	Final Concentration	Mortality (%)	Kruskal-Wallis

	LC_{50} (μg mL)$^{-1}$	(μg mL)$^{-1}$		
Control	200.00	200.00	50	-
100 mg	200.00	188.28	45	n.s
200 mg	200.00	174.36	40	n.s
500 mg	200.00	133.85	22.5	n.s

n.s. not significant.

In the FT-IR study, a fingerprint of the dye molecule was made, as shown in figure 16. It was possible to observe an intense peak at 1462 cm^{-1} referring to the stretching of the azo bond -N = N- belonging to the main chemical group of the Solar Orange 2GL P250 dye, characteristic of reactive dyes.

Another intense peak was also observed in the region of 1122-1195 cm^{-1}, these bands representing the vibration of carbon in aromatic rings.

The peaks at 1038 and 622 cm^{-1} represent the vibration and stretching of sulphonic groups which are also characteristic of the dye.

The band at 567 cm^{-1} is the -C -NH2 vibration. The band at 1632 cm^{-1} is the deformation of the -C- O band and the peak at 850 cm^{-1} represents the deformation of the C-H bonds in the aromatic rings. The peaks at 2359 and 3433 cm^{-1} are residues of carbon and primary amines that were possibly present in the environment where the readings were taken.

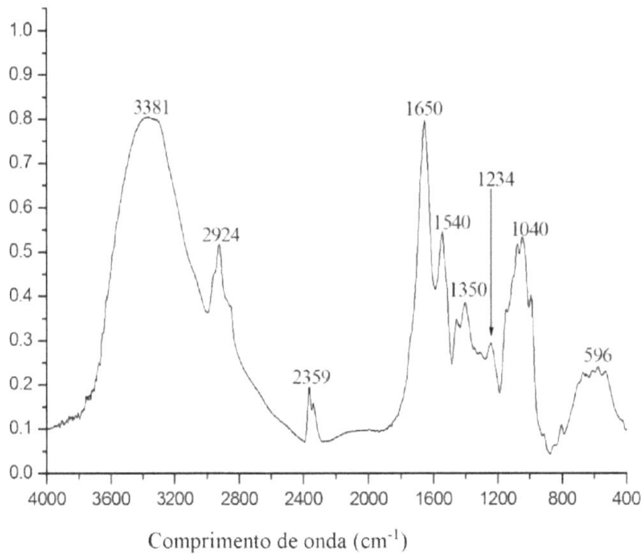

Figure 16. FT-IR of the Solar Orange 2GL P250 dye molecule.

The spectrum of the *S. cerevisiae* adsorbent (Figure 17) showed small changes in the peaks 1350, 1234 and 805 cm^{-1} which are reductions in intensity and slight deformations that are not characterized as a change in the molecule.

The 1145 and 1024 cm regions^{-1} are peaks with changes in the spectra after adsorption at pH 2.50.

These peaks are vibrations of the -C = C- bonds in aromatic rings and vibrations of -SO$_3$ groups, which belong to the Solar Orange 2GL P250 dye. The peak at 955 cm^{-1} is characteristic of sugars present in yeast cell walls, which are chemically interacting with the dye.

The peak at 608 cm^{-1} represents vibrations of -CH-NH- and -CH-NH2-, which are amine groups in yeast that are most likely protonated at acidic pH, allowing them to form another site of interaction with the dye.

The band in the region of 434 cm^{-1} is the stretch of bond between -CH - C and -C-C-C molecules belonging to the dye or some structure of the yeast cell.

It can be said that there is a strong chemical interaction between *S. cerevisiae* and the dye at acidic pH.

This corroborates the data obtained from the adsorption isotherms, showing that at a more acidic pH there is a better interaction between adsorbate and adsorbent. The yeast spectra showed no change after adsorption at pH 8.50.

This data confirms the studies of adsorption isotherms: when the pH value increases (alkaline), the interactions between the adsorbate and the adsorbent decrease due to the interaction groups available in the cell wall.

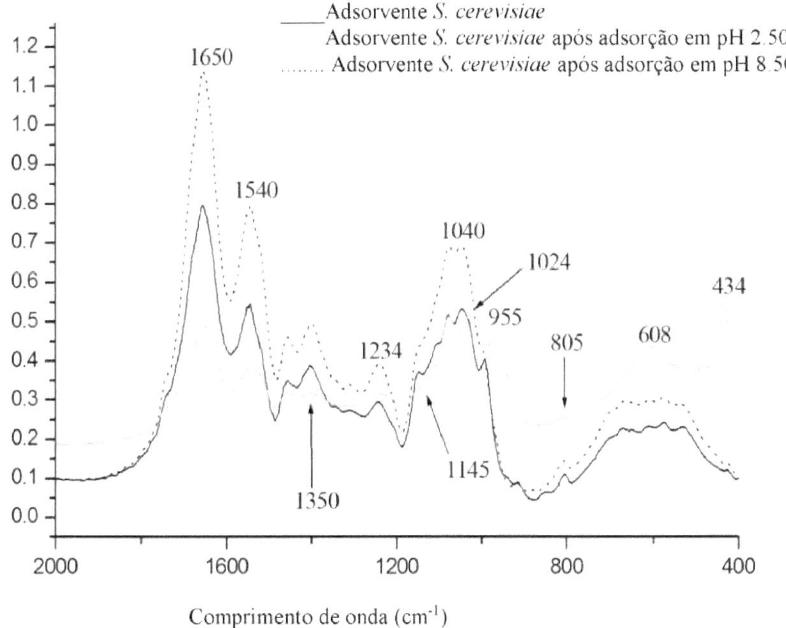

Figure 17.   FT-IR of the adsorbent and after adsorption at different pH.

CONCLUSION

Biosorption and biodegradation studies using *S. cerevisiae* biomass in a pilot bioreactor system showed the yeast's potential for treating textile industry wastewater contaminated with the reactive dye Solar Orange 2GL P250.

The use of *S. cerevisiae* biomass allows applications in bioprocesses for the treatment of textile waste, by the method of biosorption or biodegradation of dyes in textile effluents.

The yeast biomass not only shows efficiency in treating the dye, but also stands out for its ease of decanting the biomass, so the treated effluent is discarded without the *S. cerevisiae* biomass.

In addition, yeast is not pathogenic and the cost benefit is low due to its widespread use in various industrial sectors.

The biosorption studies showed performance in removing the dye in solution over a short period of time (maximum treatment time of 90 minutes) that is feasible for the textile industry, with removals of up to 70% at pH 6 (pH characteristic of the dye in solution) even at low biomass concentrations.

However, treatment by biosorption removes the textile waste from a liquid to a solid state, which reduces the environmental impact on water systems, although it does lead to the formation of new industrial waste.

A significant difference in biosorption was observed when the pH value was varied.

This is due to the interaction of ionic bonds or hydrogen bridge bonds, which can increase or decrease the number of active sites directly proportional to the acidification of the solution.

The biodegradation study makes it impossible to quantify the decolorization of the dye remaining in the solution by spectrophotometry, due to the alteration of the initial molecule, which generates various changes in the spectral structure and alterations in the chromophore groups and thus modifies the color.

Furthermore, the long treatment time makes it unfeasible for the textile industry (maximum treatment time 72 hours).

In this way, the study shows the potential for biosorption of *S. cerevisiae* for the treatment of textile effluents contaminated with the reactive dye Solar Orange 2GL P250.

As long as the pH values of the effluent containing the microorganism are controlled and the biomass concentrations are adjusted to further reduce the absorbance of the solution containing the waste.

The FT-IR studies confirmed the adsorbate/adsorbent interactions according to the changes in the spectrum peaks. At acidic pH values, the interactions are more intense than at alkaline pH. This result confirms the adsorption isotherm studies.

In the toxicity studies, the results showed that the yeast biomass reduced the toxicity of the dye in the solution. Despite the lower mortality of microcrustaceans after removing the dye, this reduction was not statistically significant.

REFERENCES

AKSU, Z., TEZER, S., Equilibrium and kinetic modeling of biosorption of Remazol Black B by Rhizopus arrhizus in a batch system: effect of temperature, *Process Bioch,* v. 36, p. 431-439 2000.

BARALDI, P. T.; CORRÊA, A. G. The use of bread yeast, Saccharomyces cerevisiae, in pheromone synthesis. *Quimica Nova.* V. 27, n.3 p.421-431, 2004.

BRUNELLI, T. F. T.; GUARALDO, T. T.; PASCHOAL, F. M. M.; ZANONI, M. V. B. Photoelectrochemical degradation of dyes dispersed in textile effluent using Ti/TiO2 photoanodes. *Quimica Nova,* v. 32, n. 1, p. 67-71, 2009.

CATANHO, M.; MALPASS, G. R. P.; MOTHEO, A. J. *Evaluation of Electrochemical and Photoelectrochemical Treatments in the Degradation of Textile Dyes.* Quimica Nova, v. 29, p. 983-989, 2006.

CONTRERAS, L.; SEPULVEDA, L.; PALMA, C. Valorization of Agro-industrial Wastes as Biosorbent for the Removal of Textile Dyes from Aqueous Solutions. *International Journal of Chemical Engineering,* v. 679352, p. 9, 2012.

COOPER, P. Removing color from dyehouse waste waters - a critical review of technology available. *Journal of the Society of Dyes and Colourists,* v. 109, n. 3, p. 97 - 100, 1993.

CHACKO, J. T.; SUBRAMANIAM, K. Enzymatic degradation of azo dye- A review. *International Journal of Environmental Sciences.* V.1, n.6, p.1250-1260,

2011.

DALLAGO, R. M.; SMANIOTTO, A.; OLIVEIRA,L. C. A. Solid tannery waste as adsorbents for the removal of dyes in aqueous media. *Quimica Nova*, v. 28, p. 433-437, 2005.

DANESHVAR, N.; KHATAEE, A. R.; RASOULIFARD, M. H.; POURHASSAN, M. Biodegradation of dye solution containing Malachite Green: Optimization of effective parameters using Taguchi method. Journal of Hazardous Materials, v. 143, p. 214-219, 2007.

DELAMATRICE, P. M. biodegradation and toxicity of textile dyes and effluents from the wastewater treatment plant in Americana-SP. *PhD thesis in Agroecosystem Ecology* - USP, 2005.

DOS SANTOS, J. R. A.; GUSMÂO, N. B.; GOUVEIA, E. R. Selection of an industrial strain of Saccharomyces cerevisiae with potential performance for ethanol production under adverse temperature and agitation conditions. *Revista Brasileira de Produtos Agroindustriais*, v. 12, n.1, p.75-80, 2010.

EL-KHAIARY, M. I.; GAD, F. A.; MAHMOUD, M. S.; SAMY, H. E. Adsorption of methylene blue from aqueous solution by chemically treated water hyacinth. *Toxicological and Environmental Chemistry*, v. 91, n. 6, p. 1079-1094, 2009.

FAVERE, V. T.; RIELLA, H. G.; ROSA, S. N-(2-hydroxyl) propyl-3-trimethyl ammonium chloride chitosan as an adsorbent of reactive dyes in aqueous solution.

Quimica Nova, v. 33, p. 1476-1481, 2010.

FLEET, G. H. Composition and structure of yeast cell walls. *Current Topics in Medical Mycology*, v. 1, n. 1, p. 24-24, 1985.

GALINDO, C. and KALT, A. UV/H2O2 oxidation of azodyes in aqueous media: evidence of a structure - degradability relationship. *Dyes and Pigments*, v.42, p.199207, 1999.

GANODERMAIERI, G., CENNAMO, G., SANNIA, G., Remazol Brilliant Blue R decolourization by the fungus *Pleurotus ostreatus* and its oxidative enzymatic system. *Enzyme and Microbial Technology*, v. 36, p. 17-24, 2005.

GOSZCZYNSKI, S.; PASZCZYNSKI, A.; PASTI-GRIGSBY, M. B.; CRAWFORD, R. L.; CRAWFORD, D. L. New Pathway for degradation of sulfonated azo dyes by microbial peroxidases of *Phanerochaete Chrysosporium* and *Streptomyces Chromofucus. Journal of Bacteriology,* v.176, n.5, p. 1339-1347, 1994.

GONG, R.; ZHANG, X.; LIU, H.; SUN, Y.; LIU, B. Uptake of cationic dyes from aqueous solu tion by biosorption onto granular kohlrabi peel. *Bioresource Technology*, V. 98, p. 1319-1323, 2007.

GUARATINI, C. C. I.; ZANONI, M. V. B. Textile dyes. *Quimica Nova,* v. 23, n. 1, p. 71-78, 2000.

HA, C. H.; LIM, K. H.; KIM, Y. T.; LIM, S. T.; KIM, C. W.; CHANG, H. I. Analysis of alkali- soluble glucan produced by Saccharomyces cerevisiae wild-type and mutants. *Applied Microbiology and Biotechnology*, v. 58, n. 3, p. 370-377, 2002.

HASSAN, A. F.; ABDELMOHSEN, A. M.; FOUDA, M. M. G. Comparative study of calcium alginate, activated carbon, and their composite beads on methylene blue adsorption. *Carbohydrate Polymers*, v. 102, p.192-198, 2014.

HOSSEINI, S. D.; ASGHARI, F. S.; YOSHIDA, H. Decomposition and discoloration of synthetic dyes using hot/liquid (subcritical) water. *Water Research*, v. 44, p. 1900-1908, 2010.

JIANLONG, W.; XIMMIN, Z.; DECAI, D.; DING, Z. Bioadsorption of lead(II) from aqueous solution by fungal biomass of *Aspergillus niger*. Journal *Biotechnology*, *v.87*, p. 273-277, 2001.

KACAR, Y.; ARPA, C.; TAN, S.; DENIZLI, A.; GENC, O.; ARICA, M. Y. Biosorption of Hg(II) and Cd(II) from aqueous solution: Comparison of biosorptive capacity of alginate and immobilized live and heat inactivated Phanerochaete chrysosporium. *Process Biochem.*, 37, n. 6, p. 601-610, 2002.

KAMIDA, H. M.; DURRANT, L. R. Biodegradation of textile effluent by *Pleurotus sajor-caju. QuimicaNova,* v. 28, n. 4, p. 629-632, 2005.

KAUSHIK, P., MALIK, A., Fungal dye decolourization: Recent advances and future potential. *Environment International, v.* 35, p. 127-141, 2009.

KIMURA, I. Y.; FAVERE, V. T.; LARANJEIRA, M. C. M.; JOSUÉ, A.; NASCIMENTO, A. Evaluation of the adsorption capacity of the reactive dye orange

16 by chitosan. *Acta Scientiarum*, v. 22, n. 5, p. 1161-1166, 2000.

KUNZ, A.; PERALTA-ZAMORA, P.; MORAES, S. G.; DURAN, N. New trends in the treatment of textile effluents. *Quimica Nova*, v. *25*, n. 1, p.78-82, 2002.

MITTER, E. K.; SANTOS, G. C.; ALMEIDA,E. J. R.; MORÂO, L. G.; RODRIGUES, H. D. P.; CORSO, C. R. Analysis of acid Alizarin Violet N Dye Removal Using Sugarcane bagasse as adsorbent. *Water, Air and Soil Pollution*, v. 223, n. 2, p. 765-770, 2012.

MÓDENES, A. N.; ESPINOZA, F. R.; ALFLEN, V. L.; COLOMBO, A.; BORBA, C. E. Use of the macrophyte *Egeria densa* in the biosorption of the reactive dye 5G. *Engevista* , v. 13, n. 3, 2011.

OGUGBUE, C. J.; SAWIDIS, T. Bioremediation and detoxification of synthetic wastewater containing triarylmathane dyes by *Aromonas hydrophila* isolated from industrial effluents. *Biotechnology research international*, v. 2011, p.1-11, 2011.

PRIGIONE, V., TIGINI, V., PEZZELLA, C., ANASTASI, A., SANNIA, G., VARESE, G.C., Decolourization and detoxi fication of textile effluents by fungal biosorption. *Water Research* v. 42, p. 2911-2920, 2008.

RAJAGURU, P.; KALAISELVI, K.; PALANIVEL, M.;SUBBURAM, V.; Biodegradation of azo dyes in a sequential anaerobic-aerobic system. *Applied microbiology and biotechnology*, v. 54, p. 268 - 273, 200.

REVANKAR, M.S., LELE, S.S., Synthetic dye decolorization by *Ganoderma sp.*

WR-1, Bioresourse Technology, v.98, p. 775-780, 2007.

SALEHIZADEH, H.; SHOJAOSADATI, S. A. Removal of metal ions from aqueous solution by polysaccharide produced from *Bacillus firmus. Water Research*, v. 37, p. 4231, 2003.

SALLES, P. T. F.; PELEGRINI, N. N. B.; PELEGRINI, R. T. Electrochemical treatment of industrial effluent containing reactive dyes. *Engenharia Ambiental*, v. 3, n. 2, p. 25-40, 2006.

SANTOS, P. K.; FERNANDES, K. C.; FARIA, L. A.; FREITAS, A. C.; SILVA, L. M. Decolorization and degradation of the red azo dye GRLX-220 by ozonation. *Quimica Nova*, v. 34, n. 8, p. 1315-1322, 2011.

SUMATHI, S., MANJU, BS., Uptake of reactive textile dyes by Aspergillus foetidus. *Enzyme Microbiology Technology*, v. 27, p. 347-355, 2000.

VITOR, V; CORSO, C. R. Decolorization of textile dye by Candida albicans isolated from industrial effluents. Journal of industrial microbiology and biotechnology, v.35, n.11, p. 1353-1357, 2008.

VOLESKY, B.; HOLAN, Z. R. Biosorption of Heavy Metals. *Biotechnology*, v. 11, p. 235-250, 1995.

YESILADA, O., ASMA, D., CING, S., Decolorization of textile dyes by fungal pellets. *Process Biochemistry*, v. 38, p. 933-938, 2003.

ZHU, M.; MENG, D.; WANG, C.; DI, J.; DIAO, G. Degradation of methylene blue with H2O2 over a cupricoxide nanosheet catalyst. *Chinese Journal of Catalysis*, v. 34, p. 2125-2129, 2013.

Printed by Books on Demand GmbH, Norderstedt / Germany